SAE EDGE™ RESEARCH REPORT

Unsettled Topics Concerning the Field Testing of Automated Driving Systems

Bob McQueen
Editor
CEO, Bob McQueen and Associates

EDGE DEVELOPMENT TEAM

Randy Iwasaki, *Executive Director, Contra Costa Transportation*

Doug Jamison, *Intelligent Transportation Systems Manager, Lynx, Central Florida Transit Agency*

Vik Bhide, *Smart City Manager, City of Tampa*

Richard Bishop, *Automated Vehicle Consultant*

Peter Thompson, *Automated Vehicle Manager, San Diego Association of Governments*

SAE INTERNATIONAL®
Warrendale, Pennsylvania, USA

About the Publisher

SAE International® is a global association of more than 128,000 engineers and related technical experts in the aerospace, automotive and commercial-vehicle industries. Our core competencies are lifelong learning and voluntary consensus standards development. Visit sae.org

SAE EDGE™ Research Report Disclaimer

SAE EDGE™ Research Reports focus on topics that are dynamic, in which knowledge is incomplete, and which have yet to be standardized. They represent the collective wisdom of a group of experts and serve as a practical guide to the reader in understanding unsettled subject matter. They are not meant to provide a recommended practice or protocol. The experts engaged have contributed their own thoughts and points of view and these are not the positions of the institutions or businesses with which they are affiliated. A professional writer has collectivized their input; there is no one contributor's perspective being advanced, but rather that of a community of practitioners. SAE EDGE™ Research Reports are the property of SAE International and SAE alone is responsible for their content.

About This Publication

SAE EDGE™ Research Reports provide state-of-the-art and state-of-the-industry examinations of the most significant topics in mobility engineering. SAE EDGE™ contributors are experts from research, academia, and industry who have come together to explore and define the most critical advancements, challenges, and future direction in areas such as vehicle automation, unmanned aircraft, Internet of Things and connectivity, cybersecurity, advanced propulsion, and advanced manufacturing.

Related Resources

SAE MOBILUS® Automated & Connected Knowledge Hub
http://sae.org/Mobilus/Automated

SAE Team

Frank Menchaca, Chief Product Officer
Michael Thompson, Director, Standards, Information and Research Publications
Monica Nogueira, Acquisitions Director
Jill Leonard, Product Manager
William Kucinski, Managing Technical Editor

Printed in USA

ISSN 2640-3536
e-ISSN 2640-3544
ISBN 978-1-4686-0121-3

To purchase bulk quantities, please contact: SAE Customer Service

E-mail: CustomerService@sae.org
Phone: 877-606-7323 (*inside USA and Canada*)
724-776-4970 (*outside USA*)
Fax: 724-776-0790

Visit the SAE International Bookstore at books.sae.org

About the Editor

Bob McQueen is an internationally recognized expert in the application of advanced technology to transportation. He has provided expert-level consulting advice to central and local governments in the Middle East, Asia-Pacific, Europe, and North America. He has also advised global companies when entering or improving positioning in the market for advanced transportation solutions. McQueen specializes in the concise communication of complex technology concepts, matching user needs to technology capabilities and constraints. McQueen has had the privilege of gaining in-depth experience in technology planning, interactive requirements analysis, international standardization, and the application of big data and analytics. He is Chief Executive Officer for Bob McQueen and Associates, based in Orlando, Florida.

contents

Unsettled Topics Concerning the Field Testing of Automated Driving Systems

Abstract

Automated driving systems (ADS) have the potential to revolutionize transportation. Through the automation of driver functions in the application of advanced technology within the vehicle, significant improvements can be made to safety, efficiency, user experience, and the preservation of the environment. According to the US Department of Transportation [1], there are more than 1,400 cars, trucks, buses, and other vehicles being tested by more than 80 companies across the USA. Implementation of ADS technology is well advanced, with many sites across the USA incorporating automated vehicles (AVs) into wider programs to apply advanced technology to transportation. Discussions with the public sector's implementing agencies suggest that one of the barriers to faster progress lies in the lack of consistent and standardized field-testing protocols. This report looks at the state of the art of field testing for ADS and identifies areas for improved consistency and standardization. It will define the problem to be addressed by AVs and the challenges associated with the introduction of such vehicles and open-road situations. In particular, the report will look at the possibilities for big data and analytics to enable the sharing of lessons learned and convergence on standard field-testing approaches.

NOTE: SAE EDGE™ Research Reports are intended to identify and illuminate key issues in emerging, but still unsettled, technologies of interest to the mobility industry. The goal of SAE EDGE™ Research Reports is to stimulate discussion and work in the hope of promoting and speeding resolution of identified issues. SAE EDGE™ Research Reports are not intended to resolve the issues they identify or close any topic to further scrutiny.

BOB MCQUEEN
CEO, Bob McQueen and Associates

Edge Development Team

Randy Iwasaki, *Executive Director, Contra Costa Transportation*
Doug Jamison, *Intelligent Transportation Systems Manager, Lynx, Central Florida Transit Agency*
Vik Bhide, *Smart City Manager, City of Tampa*
Richard Bishop, *Automated Vehicle Consultant*
Peter Thompson, *Automated Vehicle Manager, San Diego Association of Governments*

ISSN 2640-3536

Introduction

State of the Industry

ADS technology has the potential to revolutionize transportation around the globe. Significant improvements can be made in safety, efficiency, user experience, and environmental preservation. The application of ADS technology is appropriate for a range of vehicle types including private cars, transit, and freight vehicles. While AVs are still an emerging technology, there is significant activity around the USA in deploying test fleets of AVs. Discussions with public sector practitioners involved in these deployments indicate that there is a lack of consistency and standardization in the testing required for AV implementation.

This testing includes sensors that provide the perception required for AV operation, sensor fusion to combine the data from different sensors into a single picture, decision-making algorithms for threat assessment, and decision support and actuators for vehicle control and interaction with the driver. Each of these components can be tested individually in live, open-road conditions as well as the operation and performance of the entire system. Most of the deployments of ADS-equipped vehicles are implemented using test-and-validation regimes that are proprietary to the system suppliers, and consequently, information is not shared across the industry. This report addresses the need for consistent and standardized testing procedures for each aspect of automated driving and the total combined system. It also addresses the need to integrate ADS field tests within the wider spectrum of testing including simulation, test track, road tests, and full commercial introduction.

Unsettled Domains Concerning the Field Testing of ADS

The subject of ADS field testing can be approached from several different perspectives:

- By vehicle type
- By level of automation
- By the different components that comprise an ADS

There is a wide range of different vehicle types currently on the road:

- Private cars, including light pickup trucks and sport utility vehicles
- Freight vehicles, large trucks
- Transit vehicles, small and large transit buses

Vehicle Type

Private cars are assumed to be owned by an individual, while freight and transit vehicles are operated as fleet vehicles. Freight vehicles can be both long-haul, inter-city, or local-delivery vehicles within an urban environment. Transit vehicles can be the traditional 40- or 50-seat buses - either for inter-city or rapid transit solutions - or smaller local distributor-type vehicles. This category also includes "first mile, last mile" shuttles which are small 6- to 12-passenger vehicles that provide point-to-point transportation and access to and from higher density transit facilities.

Levels of Automation

SAE International has defined five levels of automation that act as a roadmap for progression from no automation to full automation (Figure 1).

This incremental approach to fully automated operation is widely accepted as a standard roadmap for current development and implementation of ADS technology. For the purposes of this report, the focus will be on Level 4 (L4) and Level 5 (L5), which involve field testing of essentially driverless vehicles.

Elements of an ADS It is generally accepted that the major components of an ADS are as illustrated in Figure 2.

Sensors. Sensors include the combination of devices that provide input to the vehicle software, including LiDAR, radar, vehicle positioning sensors, and video, among other sensor types.

Actuators. Actuators carry out the instructions of the software and include steering controls, braking controls, and accelerator controls.

Communications. There are multiple forms of communication systems including a graphical user interface to communicate with the driver and wireless communications to communicate with roadside infrastructure and other vehicles. This latter type of communication is known as Vehicle-to-Everything (V2X).

Processing Computer processors are used to receive input from the sensors, conduct multiple software functions, and instruct the actuators to act.

- **Perception** makes sense of the various data inputs from the sensors and creates a model of the vehicle, the environment, and prevailing traffic conditions.
- **Planning** - both short and long term - is carried out by making use of the model of the vehicle and environment to conduct pre-processing and make

FIGURE 1. SAE International's J3016 standard provides a taxonomy with detailed definitions for six levels of driving automation, ranging from no driving automation (Level 0) to full driving automation (Level 5). These level definitions, along with additional supporting terms and definitions provided herein, can be used to describe the full range of driving automation features equipped on motor vehicles.

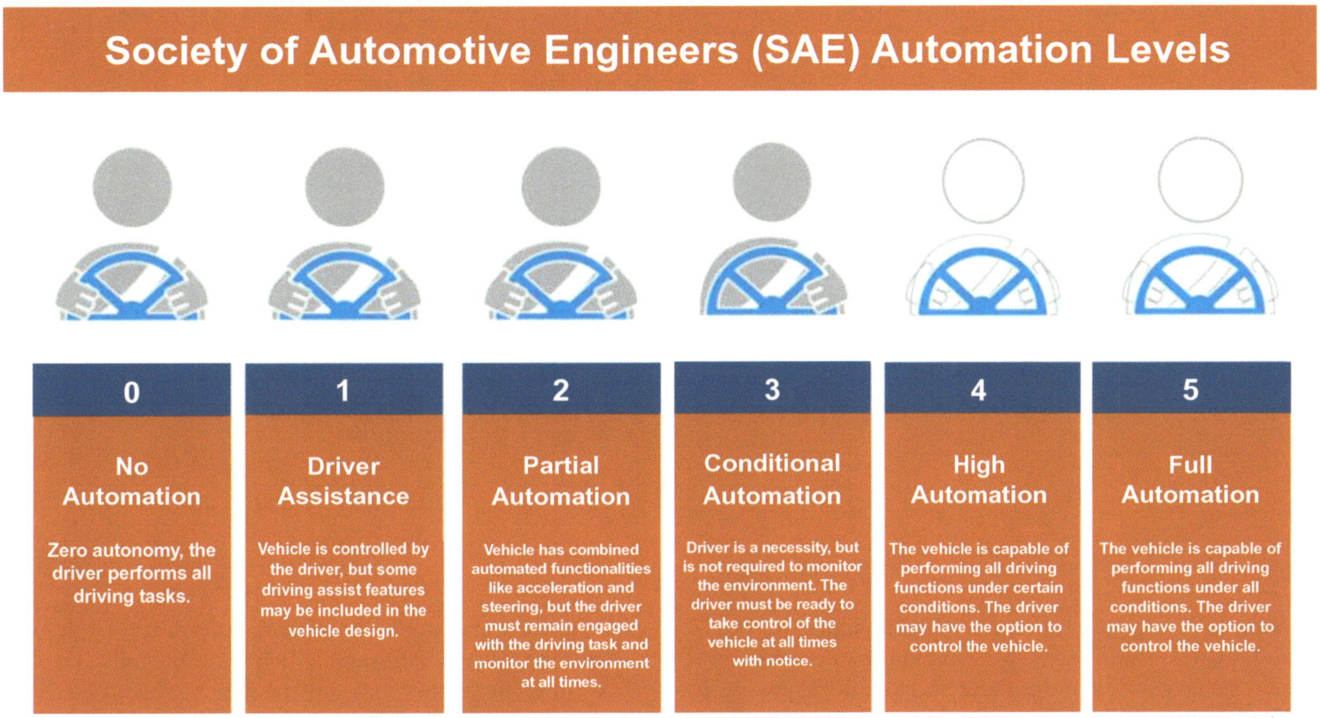

predictions. Planning is valuable to ensure that processing speed available and processing capacity is used effectively.

- **Control** defines the instructions to actuators to carry out relevant activities related to current driving conditions.

Previous SAE reports addressed issues related to the testing of sensors and the planning and execution of test track activities associated with AVs. Therefore, this report will not revisit the subjects but will build on these previous findings to discuss the implications of ADS technology from different vehicle-type perspectives. The objective will be to capture the desired output from field testing of ADS technology, from each vehicle-type perspective. It is assumed that the focus of ADS field tests will be on L4 and L5 vehicles. It is also assumed that field testing will focus on the overall performance of entire AV systems and not the performance of individual components and elements.

While further insight will undoubtedly be gained with respect to the performance of individual elements under real-world field-testing conditions, this report is written from the perspective of total-system field testing, illuminating the needs of the public and private sector from the perspectives of different vehicle types. The focus of ADS platforms is vehicles that are being tested in their entirety under real-world, field conditions.

Private Cars

ADS technology for private cars has the potential to be the most important and make the most impact on transportation worldwide. It is expected that the global market for AVs will grow from 6.6 million units in 2017 to 67.5 million units by 2028 [2]. The largest proportion of this growth will be fueled by private cars, if light pickup trucks and sport utility vehicles fall into this category. Large trucks account for approximately four percent of all registered vehicles [3]. The sheer scale of private car manufacturing and operation across the globe provides some specific challenges that must be addressed in field testing.

Scalability

Scalability becomes a crucial factor in AV field testing, from a private car perspective. Automotive manufacturers and their original equipment manufacturers (OEMs) face

FIGURE 2. Elements of an ADS.

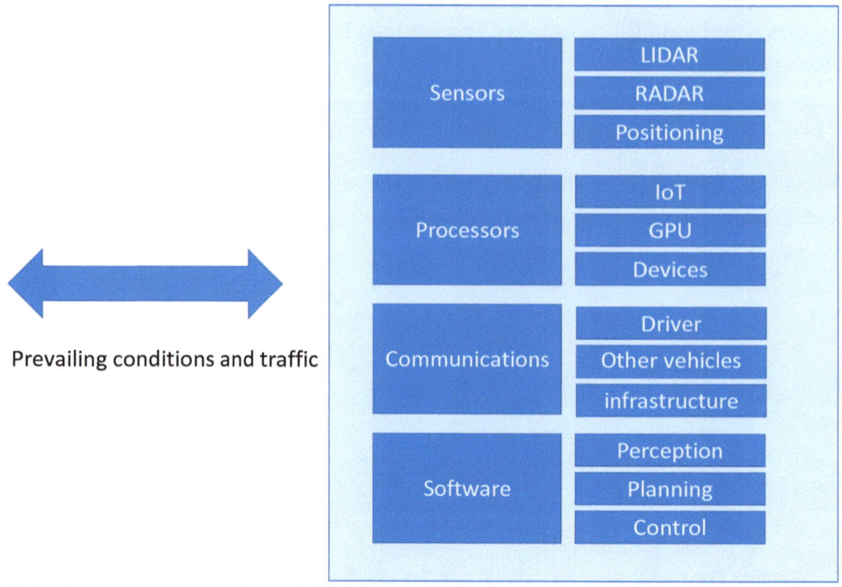

a continual challenge of introducing innovation and new technology while at the same time managing the risks associated with large-scale manufacturing and deployment. It could be argued that robust technologies implemented on a large scale with minimum risk are central to the success of automotive manufacturing. This places a significant burden on field testing to ensure that innovation and technology are as strong and understood as possible before production is scaled up to commercial levels.

Public Trust

It is essential in the course of field testing that vehicle manufacturers develop a high level of trust with the end user (i.e., vehicle occupants). This makes the implementation of field testing a delicate balancing act between fully understanding the operation of the vehicle within an open-road environment and the need to manage risk. A "high-wire" situation exists within which public reaction to incidents or accidents associated with the testing of ADS technology could be disproportionate. The need to maintain this balance and carefully manage the results from field tests makes it more difficult to establish data-sharing mechanisms to accelerate the accumulation of knowledge.

Maintenance

Establishment and application of the appropriate maintenance procedures and quality management becomes an important factor for large-scale deployment of ADS technology. Due to the limited size of tests fleets employed in field tests, it may not be possible to ascertain the full extent of maintenance needs and challenges at the field test stage. The complete picture may only become visible in the course of full-scale implementation. This makes it very important to gather all knowledge gained from the different stages of ADS testing, from the laboratories to the test track and field tests. Figure 3 illustrates the sequence of activities from laboratories through test track and road test to full commercial introduction.

Global Consistency

The focused nature of current automated-driving field tests, with specific emphasis on single companies working with single public-sector entities, does not support a global approach to field-test planning, implementation, and results evaluation. Since the automotive industry is a global business, a global level of consistency is required to ensure that ADS-equipped vehicles operate consistently regardless of country or state. This approach to ADS field testing could also have a significant impact on accelerating knowledge accumulation across the industry and across the globe.

Transition from Product Focus to Service Focus

The introduction of ADS technology comes at a time when the automotive industry is transitioning from a singular focus on the development and delivery of products to the development and delivery of associated services. For

FIGURE 3. **The AV testing spectrum of activities.**

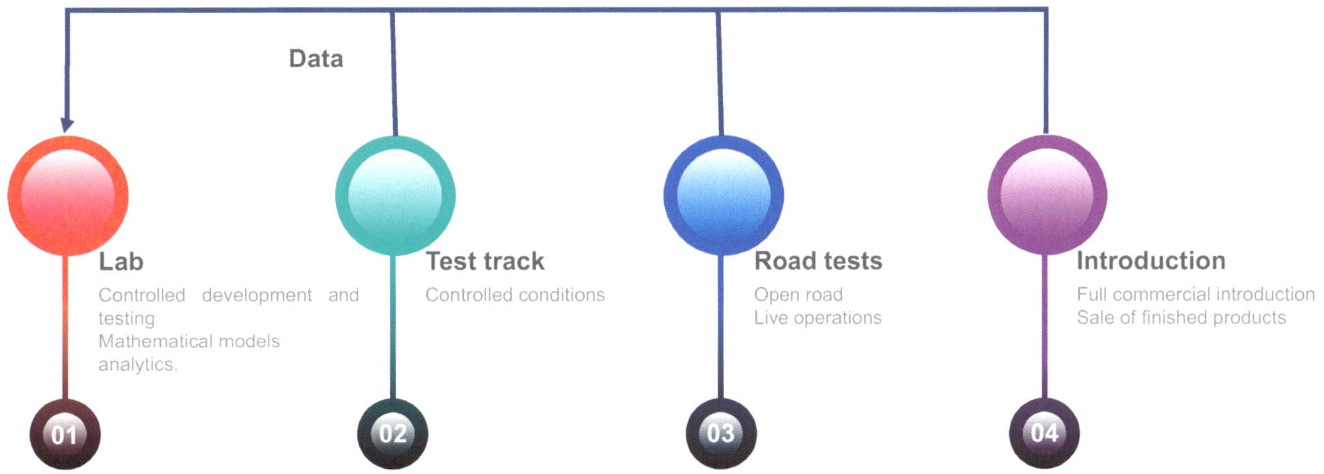

example, there is a significant focus on the establishment and operation of "mobility-as-a-service" type applications. These involve the delivery of a range of transportation services on an on-demand basis as an alternative to the purchase of a vehicle. Such services can include public transit services, private vehicles, and "first mile, last mile" shuttles.

Safety

It could be argued that the current perception related to the safety of ADS technology is unfair, with perfection defined as the only acceptable goal. Significant gains in safety can be achieved when ADS technology is better than the human manual alternative. Delaying the deployment of ADS technology could be seen as delaying the implementation of new safety technologies.

Recommendations

ADS developers should establish a framework for the accelerated accumulation of knowledge with respect to the lessons learned and experiences gained in support of safe and effective scalability of field test results. This would also include the establishment of a central repository for shareable field test data.

There is a need for awareness information that can be shared with the general public and with the public sector. The extremely focused approach taken to ADS field testing means that awareness information is limited and not consistent. The awareness information should provide a concise exposition of the value and benefits of ADS technology and address the benefits of early implementation versus a delay for perfection. Transparency with respect to the safety effects of ADS technology will be an important aspect of this material. While informational resources should include a balanced approach to the advantages and disadvantages, there is a need to provide decision-quality information to enable travelers to make informed decisions. This also extends to the public sector's technical and policy staff.

Maintenance procedures should be developed for both fleet applications and individual driver ownership. Policy development should address the full spectrum of policy needs, including the definition of liability and any compliance policies required to ensure the safety of ADS users and the general public.

This has been partially addressed by the recommendations defined above; however, a focus on global guidelines for AV operation and the interface between the vehicle and the passenger should be considered. This can also be enabled by the information sharing and central repository recommendation described above. Guidelines should address both product and service delivery as the industry migrates from a product focus to the delivery of services.

Transit

Driver and Traveler Behavior

Inappropriate use of technology can result in unanticipated and undesirable side effects. There are concerns over the introduction of connected vehicle technology that transmits signal timing information to the driver inside the vehicle. This information tells the driver when the traffic signal will change and could contribute to inappropriate driving behavior (i.e., gaming traffic interaction situations). The concern is that AVs will be treated differently than manually driven vehicles, as drivers assume that AVs will follow rules and regulations and comply in every situation, allowing drivers of manually operated vehicles to take advantage. There is a concern that the intended consequences of better information to certain vehicles could support inappropriate driver behavior.

Equity

Transit systems also have a duty to provide equitable access to transit services for all citizens. This is reinforced by the Americans with Disabilities Act [4] which does not anticipate the use of transit vehicles without drivers. For example, there are currently no rules or regulations regarding the height of the vehicle interface. While lessons can be learned from other technologies such as automated teller machines, where the height and the forward reach limits are established, there are no specific regulations for transit vehicles using ADS technology. Driverless vehicles will also require that certain facilities and features within the vehicle are redesigned to account for functions that were previously addressed by the driver or operator. This point is highlighted by the fact that most agencies around the country refer to bus drivers as operators, indicating that they have roles and responsibilities beyond driving the vehicle. There is a need for a nationally consistent approach to fulfilling operator functions that extend beyond the control of the vehicle.

While ADS technology currently has a focus on assuming responsibility for the driving function with respect to transit, it will also be necessary to automate other operator responsibilities such as passenger assistance, information, safety, and security. It would not be efficient or equitable for these additional arrangements to be implemented on an ad hoc basis from region to region. In order to provide the degree of equity required, national guidelines and practices will ensure that the solutions implemented are consistent nationwide. Note that current regulations also require that those passengers with disabilities and other constraints be able to travel to the same extent and under the same conditions as able-bodied people.

Transit agencies also have a requirement to introduce new technologies on an equitable basis. For example, they have to cycle new vehicles across the entire service area to ensure that an area or demographic is not advantaged or disadvantaged as a result of the introduction of new technology.

Bigger Picture Operations

The need for wider consideration of automated driving, beyond the control of the vehicle, also extends to the operation of automated transit vehicles within a larger multimodal transportation picture. Taking a multimodal view that considers the whole trip from origin to ultimate destination requires that several interchange possibilities be addressed. These include, but are not limited to, pedestrian-to-vehicle, vehicle-to-vehicle, and service-to-vehicle. Pedestrian-to-vehicle interchange can happen within the context of people leaving from home to work or work to home (first mile, last mile). Vehicle-to-vehicle interchange can occur when the traveler moves from one mode of transportation to another. This could include interchange between different transit services such as bus to bus or bus to train. Interchange for one service to another can take place when the traveler moves from a service operated by one service provider to another.

An interesting example of pedestrian-to-vehicle interchange could occur when travelers leave a large office building to make the journey home. A transit vehicle equipped with an ADS now becomes part of a larger coordinated activity to ensure that there is enough capacity for many people in a short period of time. The assignment of travelers to different vehicles must also be addressed.

Unexpected Retail Impact

It has been observed that traditional commuters using private cars may not make a direct journey home, but may stop along the way to take advantage of retail and refueling locations. If ADS technology applied to transit or private cars could remove these interim stops, there could be an unintended impact on local retailers. While this is obviously a primary concern to the private-sector retailers, it is also of concern to transit agencies that benefit from local taxation derived from retail activities.

Support Infrastructure

With respect to the incorporation of ADS technology within a bigger and coordinated picture of regional and city transportation, there is a need to ensure that ADS technology is fully integrated within a larger multimodal transportation service delivery framework. The example described above, where many people leave a large office building within a short period of time, indicates the need for management and coordination of multiple transportation modes. This will also include the provision of appropriate infrastructure to support interchange from pedestrian to vehicle and from vehicle to vehicle. Management of infrastructure availability and assignment of vehicles to appropriate pickup locations would be an obvious application for information technology and mobile apps. Ideally, the management system would have links to and would coordinate with other transportation management systems within the region.

With respect to procurement, it will be necessary for public transit agencies to determine not just the lifecycle cost of the vehicle but also the arrangements required to maintain ancillary support infrastructure, as well as vehicle technologies.

A complete systems approach to ADS technology within the larger context would also address opportunities to optimize infrastructure such as signs and road marks to better support automated driving operations. This may require revisions to the current Manual on Uniform Traffic Control Devices (MUTCD).

Digital Infrastructure

Transit agencies also have a concern with respect to the management and maintenance of the digital infrastructure required for ADS technology. This includes digital maps and arrangements for maintenance and upkeep of the data required to support ADS technology. An example quoted was the removal of a stop sign along a bus route. It is not clear who would be responsible for maintaining a database of such data and making sure that this is provided to the relevant ADS operating companies. Transit agencies are concerned with an answer to the question "Who do I tell if something goes wrong?" This concern also extends to the need for real-time updates as operating conditions for AVs change in real time. This includes road closures and emergency situations that require diversion of routes and traffic.

Concept of Operations

Especially for transit agencies, there is a strong need for a practical concept of operations to be developed regarding the use of automated transit vehicles in real-world, practical situations.

Technology Adapts to People and Operations, Not the Other Way Around

Customer-focused transit agencies are concerned that current ADS implementation places too much attention on vehicle technologies while treating passenger and operator concerns as an afterthought.

There is a preference among transit agencies for the opposite to be the case and to ensure that the AVs are designed to fit with all passenger needs and to operate within the context of existing operating procedures. Regarding passengers, this will address the full spectrum of passengers including those with disabilities. Transit agencies are also concerned that there may be division with respect to generational perception. Older passengers may have no prior exposure to new technology, while younger passengers with a strong technology background may insist on it. There is a belief that this generational gap also extends to those travelers with disabilities. New technology introduction approaches feature fail-fast implementation models that may not be appropriate for transit.

There is also a tendency to introduce ADS technology in smaller vehicles. This approach does not fit with the overall goal of adapting the technology to fit current operating procedures which may require the operation of full-size buses.

There is a strong need to keep in mind the end customer and ensure that AV designs fully address the needs of the entire customer group.

Workforce Development

Workforce development is also a concern to transit agencies. There is a need to determine what the skill set requirements will be in a future operating system area that requires the use of AVs. Transit agencies will expect that field testing will shed light on this issue. At the present time, there is a concern that the skill sets do not currently exist within transit agency organizations. There is also a perception that training and professional development activities to generate the required skill set are also not currently available, and this would be an area for development.

Technology Is Ahead of Policy

There is a strong perception that the capability of technology has moved ahead of public policy. While this is an issue for other modes of transportation, it is particularly important to transit agencies.

Recommendations

Field testing should include elements of driver and traveler behavior research and measurement. It is necessary to not only understand the performance of the vehicle in real operating conditions, but also the response of drivers and travelers. This will include assessing the effectiveness of interfaces between the vehicle and the driver and the traveler. Particularly, they should include an assessment of the full range of proposed system users including those with disabilities.

Field testing should also include consideration of equity, exploring how to deliver automated driving supported services for the full spectrum of clients. The requirements of the Americans with Disabilities Act should be recognized when developing field-test plans and implementing field tests.

Field tests should also take account of the context within which the ADS is operated. The best results will be obtained when AVs are implemented in a coordinated and harmonious fashion considering other transportation services within a city or region. Field tests should account for the coordination required between ADS technology and other services and the need for data sharing across multiple modes of transportation. Coordination of management strategies for ADS technology and other forms of transportation should also be explored in the course of field testing.

Unexpected retail impact is only one of the potential serious consequences that will result from the field testing of ADS technology. Field tests should take a wider view beyond the focus on the vehicle, and consider the economic and social effects of technology application within the bigger picture. This would require that field tests have a wider purview to examine how communities will respond from a business perspective.

Results from field tests should also include a clear definition of the support infrastructure required to correctly support the operation of the ADS technology.

While most field tests are currently conducted under the auspices of a concept of operations for the field test, it would be most beneficial if field-test results are also used to create a detailed concept of operations for commercial introduction of the technology. The concept of operations should provide an overview of how ADS technology will operate, and a detailed exposition of roles and responsibilities for both the public and private sector.

The whole development process for ADS technology, including field testing, should take full account of the characteristics of the user and put a primary focus on meeting user needs through adaptation of the technology. It is important to take a user-driven, customer-centric approach, rather than assume that the user will adapt to current technology features. This is important to the end user and to the operating agencies responsible for interfacing and managing ADS technology.

An important output from field tests will be a detailed definition of the skill sets required to support commercial introduction and operation of ADS technology. This should include consideration of support needs for private vehicle operation and fleet operation. It is important to develop a clear picture of current skills, future required skills, and a detailed analysis of the gaps. The workforce development plan would also provide an overview of arrangements required to fill the gap between current and required skills.

The most effective field testing will also provide input into the policy required to support technology implementation. Practical input into the balance between compliance and innovation can be provided as a result of field-testing experience. This will require close cooperation between the public and private sectors.

Freight

The development of ADS technology for freight is at the leading edge of current field testing. There are obvious economic benefits in replacing the driver with automation for both long-haul inter-city freight and local urban logistics.

Field-testing activity is being led by several private-sector organizations, with the general perception that standards are not required at this stage of development. The view is that standards would be more appropriate at a later stage, after product introduction.

The private sector implementation has led to a somewhat fragmented approach to field testing around the country. Private sector organizations are working with public sector authorities to develop specific plans and rules for field testing. While the focus on individual sites can be extremely effective and efficient, it does not provide the basis for sharing data, combining lessons learned, and accelerating the state of the art through knowledge acquisition and application. To some extent, data and knowledge sharing is perceived as a negative as such data and knowledge is a differentiator and a competitive advantage for the early leaders.

To some extent, the safety of the public is dependent on the responsible attitude being adopted by each individual private sector organization. Each organization has a burden of proof and responsibility for each individual field test; however, all can be at risk if the activities of an irresponsible company ignore safety protocols in these early days of field testing and implementation.

The general view is that self-certification, rather than policy, regulation, and guidelines on a national level, is the best way to approach. With the whole area of activity being driven by private-sector action and investment, it is viewed that government could have difficulty in developing the appropriate documents. There is also concern that high-level government intervention could stifle innovation. On the other hand, the lack of appropriate standards and regulations could make it more likely for private-sector companies to be sued in the case of an incident.

High-Wire Effect: The Likelihood of an Extreme Public Reaction to Accidents

There is also concern regarding the appropriate awareness and understanding of the potential public sector roles in field tests. Public officials should be educated on these roles and how they play within the realm of ADS field tests. Currently, they are being defined as part of competitive actions to attract private companies to field test within specific regions. In many cases, public sector participation evolves as part of a dialogue with the private sector on specific field test possibilities. However, there is a need for more effective awareness campaigns to brief public-sector staff and officials on essential aspects of automated driving field tests.

The private-sector drive also makes it difficult to define how the bigger transportation picture within which ADS will fit. Essentially, the private sector can design the freight vehicle to be as efficient and effective as possible but, cannot guarantee that it will not get stuck in a traffic jam. More public-private cooperation is required to achieve the optimal effect.

Public-sector cooperation and alignment is also required for the development and investment in test track facilities. There appears to be a surplus of publicly-funded test-track facilities, suggesting that a greater deal of coordination and alignment is required to ensure the effective use of funds.

The relatively small sample sizes associated with current field tests can also make it difficult to precisely measure the effects and performance. The fragmented nature of field testing, with the lack of support for data and information exchange, is exacerbating the situation.

Freight is attractive as an early entry into ADS field testing as most freight is subjected to professional fleet and maintenance management.

Recommendations

To a large extent, the private sector has already addressed the development of the economic case for ADS technology and freight. However, there is a need to develop a common and agreed-upon understanding of the economic case and ensure that this is shared with the public sector.

An awareness campaign will be specifically focused on the needs for freight, and address both public and private sectors, along with end users. This will include the economic case described above and provide additional background on the anticipated impact of freight automated driving and the respective roles of the public and private groups. Roles and responsibilities for the public sector will be an important part of this activity.

Challenges Identified across All Domains

Based on the overview of the private car, transit, and freight perspectives, a number of challenges have been identified and summarized as follows:

1. The lack of appropriate standards for AV field testing
2. No arrangements available for sharing data, lessons learned, and practical experiences
3. Too difficult to guarantee the safety of the public
4. The lack of formal regulations to govern and enable AV operations on open roads
5. The lack of established policies to govern field-testing operations and enable innovation
6. High-wire effect: the likelihood of an extreme public reaction to incidents or accidents
7. The technology is not yet ready
8. The necessary roadside infrastructure and supporting arrangements for AV operation are not yet present
9. Inadequate public-sector support for the essential roles and responsibilities of the public sector
10. It is difficult to define an upward growth path to enable scaling from field testing to full commercial implementation
11. It is a challenge to relate laboratory work, test-track work, and field tests as a single process
12. It is difficult to integrate AVs within the bigger carbon-smart mobility picture
13. The lack of suitable arrangements for big data management and analytics
14. Ineffective risk management
15. Ambiguity in operations roles and responsibilities
16. Difficult to identify ideal field-test facilities
17. The lack of appropriate maintenance policy
18. Ineffective performance measurement and management
19. The lack of clearly defined outcomes and objectives
20. The public sector does not engage the private sector effectively
21. Slow public-sector business processes delay the field testing

The Lack of Appropriate Standards for AV Field Testing

There are currently no national standards for AV field testing. The field testing taking place all over the USA is guided by proprietary approaches developed for each specific site and by each of the private sector enterprises involved. Additionally, there are currently no formal arrangements in place for sharing data, lessons learned, and practical experiences. Indeed, there is a widely held view that such arrangements are not required and not wanted at this point as data collected from field testing is proprietary and a differentiator for the company involved.

Figure 4 illustrates an ideal situation where the bottom-up activity from the large number of field tests is brought together through the development of a framework for top-down knowledge sharing and encouragement.

Too Difficult to Guarantee the Safety of the Public

This may not be as big of a challenge as expected. If ADS safety levels are compared to safety levels for currently manually operated vehicles, then it is safe enough. However, if the ADS benchmark is a perfect driving record, then the technology may not yet be ready. At this time, there are no formal ways to assess the risks and determine the level of ADS technology for the public.

The Lack of Formal Regulations to Govern Field Testing and Operations on Open Roads

This is closely related to the previous challenge, as there are currently very few formal regulations to govern and enable

FIGURE 4. **Evolution of the technology-based product.**

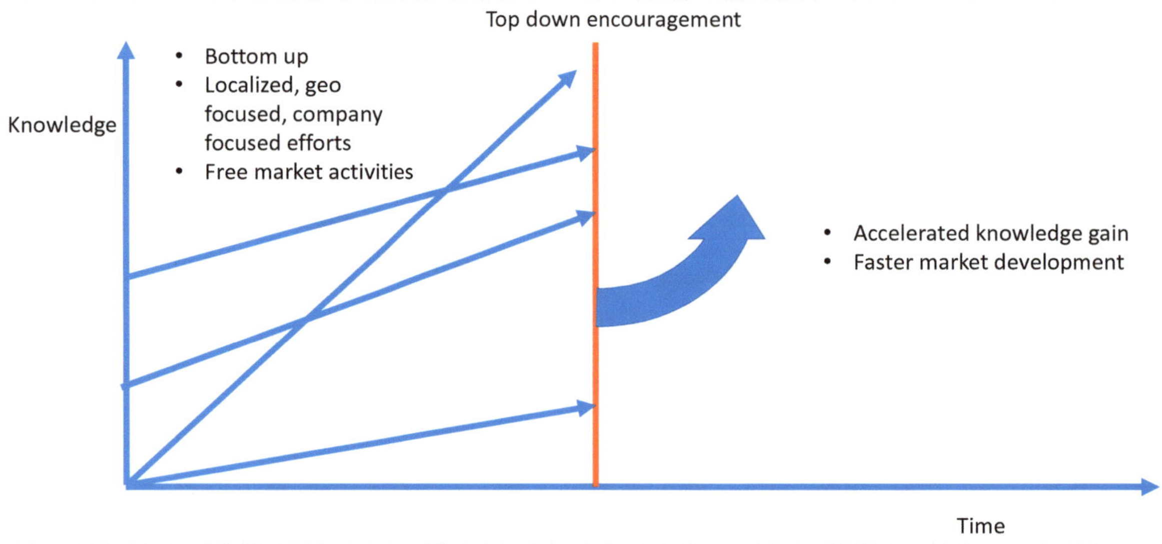

AV operations on open roads. The regulations that do exist are very specific to the city or region in which the testing is occurring, and that is not a national approach to regulation.

We have entered a situation where technology capabilities have exceeded the span of available policies. Technology and private sector investment are currently driving ADS technology into existence. There is a lack of knowledge on the public side regarding the implications of this and the roles and responsibilities of the public sector.

Technology Readiness

Some of the levels where SAE's five levels of automation can already be addressed by current ADS technology. It is likely that full L5 automation will be within the capability of the technology within two years. The ways in which the capability of the technology can be accelerated and rapidly improved could be very valuable in this respect, such as creating the necessary roadside infrastructure and supporting arrangements for AV operation.

Infrastructure and supporting arrangements are available on a city-by-city or region-by-region basis. Very often, the private sector will choose a specific city or site because of the availability of support arrangements and infrastructure. It would be fair to say that there is no nationally consistent availability of supporting infrastructure and arrangements for ADS technology.

Communicating Simulation, Test Track, and Field Work

The private sector seems to have overcome this challenge and has seamlessly connected lab, test track, and field tests. This may not be the case consistently across all private-sector organizations, as smaller startup companies may have the lack of professional business processes and know-how to achieve a seamless delivery of this process. They may have difficulty relating products developed in the laboratory by private organizations with public test tracks.

Integrating AVs into the Smart City

If the public sector is not completely aware of the possible implications of the ADS technology, then it becomes difficult to integrate this technology within the bigger smart-mobility picture. There is considerable interest in investment in smart cities, with mobility or transportation at the front end of development work. Initiatives such as mobility as a service and the emergence of privately operated microtransit systems offer new possibilities to improve accessibility and mobility within an urban area. It is vital to understand how ADS technology will fit within this new framework.

Big Data Management and Analytics

The data being collected for field tests is proprietary and is not shared from one company to another. This protects the private sector's investment but stifles the ability to gain knowledge faster and support rapid implementation of the technology. Big data management and analytics can also play a crucial role in setting performance standards relating to safety, efficiency, and user experience that would be extremely valuable to the public sector in ensuring that the technology is harnessed successfully.

FIGURE 5. **Public sector approach plan.**

Ineffective Risk Management

While each project has its own risk management arrangements and, in most cases, these are implemented professionally to risk-management best practices, there are no arrangements to ensure consistency across the multiple field tests being conducted at the current time.

Ambiguity in Operations Roles, and Responsibilities

While the private sector is very clear with respect to operations roles and responsibilities, the public sector is under other challenges and struggles to understand and clearly define roles and responsibilities. The public sector seeks a delicate balance between compliance for safety and the

FIGURE 6. **Actions to support the approach.**

achievement of the community optimal with respect to efficiency, while also enabling the private sector to invest and innovate.

Identifying Ideal Field Test Facilities

Field test facilities tend to be self-selecting, with the private sector gravitating toward cities and regions that are open and willing to cooperate. It could be argued that the major selection criteria for field test locations has centered on public sector willingness to allow the private sector to test. This may not provide the ideal set of circumstances and conditions for effective field testing.

Maintenance Policy

Many of the early initiatives with respect to ADS field tests relate to the operation of vehicle fleets. This avoids the issue of appropriate maintenance policies, with maintenance conducted by professional staff. This still leaves the open case of consistent and appropriate maintenance standards for private cars.

Ineffective Performance Measurement and Management

The effectiveness of performance measurement and management can be defined from two different perspectives. From the perspective of the private sector, performance measurement and management are adequate, but focused on the private sector. There is a weakness in performance measurement and management for public-sector needs related to the effect of incorporation of the technology within wider transportation service delivery in a city or region. There is a lack of national benchmarks and baselines for performance management and a strong focus on individual private organizations, cities, and regions.

Outcomes and Objectives

While outcomes and objectives are clearly defined for each city or region, there is no nationally agreed-upon set of outcomes and objectives for the introduction of ADS technology. This is a separate issue related to the overall lack of public awareness concerning ADS technology. However, increasing public-sector awareness requires establishing achievable outcomes and benefits.

Combined System Road Tests

Different solutions to automated driving make use of different approaches to combined-system testing. Most implementation tests on the road are conducted under the protocol for either the total-system supplier or the combined-system supplier. To improve safety and smooth the introduction of ADS technology, a consistent, coherent, and standardized framework for combined-system road testing is required.

Ideally, a combined-system approach to road tests or field tests would also support the integration of lessons learned from the various stages of testing including laboratories, test track, and field tests.

Field-testing AVs is only one step in a chain of events over the course of which the technology matures to the point where it can be implemented in the open road. In the laboratory, technologies and software are developed and tested under controlled conditions. On the test track, the situation moves closer to real life with less control than in the laboratory conditions, but still in a carefully managed environment. Finally, when the first two steps have been successfully completed, the technology is ready for the field. In the field, there are open-road environments where unexpected things can happen, and the operation of the ADS is not under the controlled environment of the laboratory or the test track. In the case of ADS technology, this is another possible step between test track and field, in which a large volume of observed data from manually operated vehicles in the field can be collected and used to identify a wider range of variations and permutations in the conditions that the AV will encounter in real-world driving. The process of moving the ADS from the laboratory to the test track and then to the field should be viewed as a continuum. There will be lessons learned along the way and practical experience will be gained which will contribute to collective industry knowledge.

There is also value in a feedback loop from field experiences back to the laboratory. New information received in the field regarding expected operating conditions and vehicle performance can be fed back into the laboratories to provide guidance and input to future product development. An interesting point in this is the question of who should be responsible for each stage of the process. There is an argument for different groups of people to be responsible for each stage of the process as there is a tendency for the "inventor" to protect or shelter the "invention" whether consciously or subconsciously. In practice, it is necessary for the invention to be subject to the rigorous and widest range of challenges. Therefore, it may be better for different groups of people to be responsible for management and implementation of laboratory work, test track, and field tests.

Summary/Conclusions

Field testing for ADS technology is well underway in many test sites around the country. The current programs feature fragmented and focused effort in cooperation between individual private-sector solution providers and specific public-sector agencies. There is little in the way of information sharing that could support accelerated learning and technology implementation.

There is no question that the private sector is leading the development and implementation of ADS technology, with "follower" style cooperation from the public sector. In many cases there is a lack of awareness within the public sector on appropriate roles and responsibilities for ADS field tests. The level of awareness is inconsistent and uneven across the country, with those public agencies involved in current field tests that have a higher degree of awareness and knowledge of the subject than those not involved. The private sector perceives lessons learned and information gained from field test as proprietary information that can differentiate their offering, making it difficult to encourage information sharing. The three unsettled domains identified - private car, transit, and freight ADS technology - have significant challenges as listed in the report. There is a need to address these challenges in order to ensure wise investment and efficient knowledge accumulation across the industry and across the country. While the current approach to automated driving field tests were undoubtedly with individual companies' forward motion, there is a possibility that the industry will not move as quickly as it could due to the lack of common-knowledge critical mass.

Where Do We Go from Here?

It is important to take a balanced approach to data, information, and knowledge sharing that enables individual private-sector solution providers to keep differentiating information proprietary. On the other hand, there is a growing recognition that the overall market can be accelerated, advanced, and expanded by a degree of sharing at the appropriate level and with the appropriate subject matter.

It is very important to develop a range of materials that are designed to raise awareness, especially among public sector partners, on appropriate roles and responsibilities associated with ADS field tests. Other factors that will impact the development of a long-term solution are listed below:

- A clear exposition of the economic justification for such involvement as this would be a necessary part of the overall value proposition required to convince public sector agencies to be more proactive.
- Field tests take a wider view of overall objectives beyond the performance of the vehicle within the open-road environment.
- The behavior of drivers and other travelers and how they interact with the new technology should be considered.
- A user-focused approach to the design of solutions should be employed. This will ensure that they adapt to the needs of users and operating agencies, rather than the reverse.

It is an exciting time in transportation, with considerable private sector investment in ADS technology and growing public sector interest in harnessing the benefits in terms of safety, efficiency, and enhanced user experience. While it has become fashionable to host an ADS field test, it is more important in developing a sustainable approach to this that the business case is also recognized and documented, and that forward motion is guided by an understanding of suitable outcomes. The ADS itself is not the outcome or the end objective for the work, it is merely the vehicle through which value and benefits are delivered by a range of services. These ADS services are also required to fit within a wider context of transportation services within urban and rural areas.

History shows that most innovation and technology start on a bottom-up basis in a fragmented manner, very unlike the current ADS field tests. It also shows us that the introduction of a structural framework at the appropriate time can have the effect of accelerating progress, ensuring the achievement of objectives, and avoiding undesirable side effects, while continuing to encourage and stimulate innovation. The time is just about right to consider such a structured framework that will coordinate and take full advantage of the progress made so far.

SAE EDGE™ Research Reports

SAE EDGE™ Research Reports, like the present report on "*Unsettled Topics Concerning the Field Testing of Automated Driving Systems*," are intended to push further out into still unsettled areas of technology of interest to the mobility industry. SAE launches these reports before attempting to form a joint working group, let alone a cooperative research program (CRP) or a Standards Committee.

SAE EDGE™ reports are intended to be quick, concise overviews of major unsettled areas where vital new technologies are emerging. An unsettled area is characterized more by confusion and controversy than established order. Early practitioners must confront an absence of agreement. Their challenge is often not to seize the high ground but to find common ground. These scouting reports from the frontiers of investigation are intended merely to begin the process of sorting through critical issues, contributing to a better understanding of key problems, and providing helpful suggestions about possible next steps and avenues of investigation.

SAE EDGE™ Research Reports, therefore, are fundamentally distinct from the more formal working groups approach and far removed from the more mature CRP and standard's development process. They begin by convening an informal meeting of subject matter experts working in an unsettled area of technology from industry, government, and the research community. A moderated but collegial discussion ensues regarding their shared interest. This conversation attempts to identify critical issues hindering a clear understanding of important topics, preventing a common agreement on potential solutions, or slowing progress in key areas of research.

An SAE EDGE™ Research Report is then drafted and circulated for comment to meeting attendees and other interested parties.

Next Steps for ADS Field Tests

This publication should be considered only as a first step toward clarifying the issues around ADS field tests. The intention behind this and other SAE EDGE™ Research Reports is to start a dialogue among interested parties on important industry-wide topics that require further attention. The expectation is that these explorations of unsettled areas of technology will lead to the formation of working groups and, ultimately, committees that can address and resolve the issues they raise, producing a framework for developing a common vocabulary of definitions, best practices, protocols, and standards needed to support continued progress toward safer and more innovative automotive products.

The experts' discussion that gave rise to this publication demonstrated a great willingness on the part of the industry to define the terminology, procedures, and eventually the standards needed to enable ADS field test technology to move ahead as quickly and efficiently as possible. SAE International has demonstrated its lead in this and closely related areas by the SAE work regarding ADS field testing.

Recommendations

The overall recommendations of this SAE EDGE™ report can be summarized as follows.

Data and Knowledge Sharing The establishment of a data-sharing mechanism for data that emanates from AV testing. This should enable data to be accumulated from as many field tests as possible and subjected to a stringent data-governance plan that is pre-agreed with those that will supply the data. The data-sharing mechanism should be marketed to both the private and public sectors as a means of accelerating the implementation of ADS technology. The following data should be collected and brought to an accessible, central location:

- Vehicle performance data
- Field test data
- Public reaction data
- Driver behavior data
- Challenges
- Opportunities
- Best practices

Knowledge Sharing Knowledge sharing would follow the same pattern that is specified for data sharing, but provide additional capability to turn data into information, into insight, and then into knowledge. A focus of the knowledge or know-how should relate to the common knowledge that is being developed across all test sites regarding vehicle and driver behavior. It may be necessary to divide the knowledge base into that which is proprietary and not shared, and that which is common and leads to the acceleration of the implementation of ADS technology.

Public Approach Support It has been identified that there is a need to raise awareness to the public sector levels regarding the likely impacts and the benefits of ADS technology. A public-approach support program should be created that provides for the development of a range of materials, tools, and resources to support effective public involvement in AV field testing and subsequent commercial implementation. This should include the following materials at a minimum.

ADS Training Materials for Both Technical and Nontechnical Staff ADS technology impacts and benefits information for nontechnical decision-makers:

- Definition of a standard planning approach for AV testing and implementation
- Collection of best practices information into an accessible resource
- Development of a website or portal as a common access point for awareness materials
- Benefits and costs summary
- Advice on how to develop customized plans for specific cities and regions
- Advice on the development of a vision statement for automated driving within a city or region

Collaborative Mechanisms Collaboration can be supported through several channels. These can include special interest groups, workshops, conferences, and the use of social media for collaboration.

Encouragement for Innovation A range of documents should be developed that specifically focus on encouraging innovation with respect to ADS technology. These can include case studies and highlight success stories, along with the challenges and opportunities that were encountered and addressed. A major component would also be related to the expected benefits and impacts of ADS technology on mobility within urban and rural areas.

Encouragement for Partnership Here again, case studies and information to support the use of best practices for establishing and managing partnerships related to ADS technology. The information can be delivered over multiple channels including presentations at conferences, web, YouTube, and social media.

Abbreviations/Definitions

ADS - Automated Driving System. Intelligent computer control of vehicles based on sensor input, characteristic of vehicles at L3 and higher *Automated Driving Levels*. See the entry for this term below.

Automated Driving Levels (as defined by SAE J3016™)

L0 - No Driving Automation. A human driver performs the entire Dynamic Driving Task (DDT). Other vehicle systems may provide warnings or support, e.g., braking systems with automatic skid control.

L1 - Driver Assistance. The driving automation system is capable of executing either the longitudinal or the lateral vehicle motion control subtask (not both). It disengages immediately upon driver request. A human driver is present at all times to perform all control tasks not done by the automation system, to determine when it is appropriate to engage the automation system, to supervise all operations of the automation system, and to intervene as necessary to maintain safe vehicle operation. Example: automatic speed control ("cruise control") for highway driving.

L2 - Partial Driving Automation. The driving automation system is capable of executing both the lateral and the longitudinal vehicle motion control subtasks. It disengages immediately upon driver request. A human driver must be present and attentive at all times to perform control tasks not done by the automation system, to determine when it is appropriate to engage the automation system, to supervise all operations of the automation system, and to intervene as necessary to maintain safe vehicle operation. Total driver disengagement from human control mechanisms (hands off the wheel and feet off the pedals) is possible in narrowly limited driving scenarios, e.g., lane centering.

L3 - Conditional Driving Automation. A human driver must be present and attentive at all times to operate the vehicle outside the Operational Design Domain (ODD) of the ADS and to take over control of the vehicle from the ADS within its ODD as necessary. The ADS can only be engaged in its ODD but can perform the entire DDT when engaged. It automatically disengages when it determines that either its ODD limits are about to be exceeded or it is experiencing a performance-relevant system failure, first, by issuing a timely request to intervene to the human driver and, second, by disengaging an appropriate time after issuing the request. It also disengages immediately upon driver request. The human driver is responsible for verifying the operational readiness of the vehicle, handling all DDT outside the ODD of the ADS, determining when the ADS can be engaged, and serving as the fallback-ready driver when the ADS is engaged and responsible for maintaining a minimal risk condition by actively intervening in ADS operation or assuming vehicle control on automatic ADS disengagement. An ADS ODD might include, for example, freeway driving or light traffic on open roads outside urban areas.

L4 - High Driving Automation. Human driver is responsible for verifying the operational readiness of the vehicle and determining whether to engage the ADS. The ADS will permit engagement only within its ODD. When the ADS is engaged, the human driver has no fallback role and may ride as a passenger. Once engaged, the ADS will transition to a minimal risk condition before it disengages (1) as the result of a performance-relevant system failure, (2) after an automatically issued request to intervene is ignored by a human driver, or (3) in response to a user request that it achieved a minimal risk condition. Optionally, a human driver may reassume DDT by requesting that the ADS disengage and waiting for it to turn over control of the vehicle once the ADS determines that conditions are appropriate; or, in a fallback role, following an ADS request to intervene; or, after the ADS has automatically disengaged, following achievement of a minimal risk condition.

L5 - Full Driving Automation. The ADS ODD expands to permit engagement under all driver-manageable on-road conditions. L5 vehicle operation, when the ADS is engaged, is identical to L4 vehicle operation.

AV - Automated vehicle

LiDAR - Light detection and ranging

OEM - Original equipment manufacturer

MUTCD - Manual on Uniform Traffic Control Devices

V2X - Vehicle to everything

Acknowledgments

Recognition should go first to all the participants, many of who also provided feedback on the draft version of this publication. Without their input and initiative, this SAE EDGE™ Research Report would not have been possible.

1. Randy Iwasaki, Executive Director, Contra Costa Transportation
2. Doug Jamison, Intelligent Transportation Systems Manager, Lynx, Central Florida Transit Agency
3. Vik Bhide, Smart City Manager, City of Tampa, FL
4. Richard Bishop, Automated Vehicle Consultant
5. Peter Thompson, Automated Vehicle Manager, San Diego Association of Governments

Works Cited

1. U.S. Department of Transportation, "Preparing for the Future of Transportation: Automated Vehicles 3.0 (AV 3.0)," 2018, https://www.transportation.gov/av/3, accessed November 21, 2019.
2. BIS Research, "Global Autonomous Vehicle Market: Focus on Level of Autonomy for Passenger Cars and Commercial Vehicles, Analysis and Forecast: (2018-2028)," February 2019, ID: 5748373, https://www.reportlinker.com/p05748373/Global-Automated-Vehicle-Market-Focus-on-Level-of-Autonomy-for-Passenger-Cars-and-Commercial-Vehicles-Analysis-and-Forecast-.html, accessed November 21, 2019.
3. NHTSA, "Traffic Safety Facts," May 2016, https://crashstats.nhtsa.dot.gov/Api/Public/ViewPublication/812279, accessed November 21, 2019.
4. ADA National Network, "An Overview of the Americans With Disabilities Act," https://adata.org/factsheet/ADA-overview, accessed November 21, 2019.

Contact Information

EDGEresearch@sae.org.

The authors of this document together with the SAE Team responsible for its creation join in expressing our deepest appreciation to all of the individuals who contributed.

www.ingramcontent.com/pod-product-compliance
Lightning Source LLC
LaVergne TN
LVHW070942210926
848482LV00003B/280
* 9 7 8 1 4 6 8 6 0 1 2 1 3 *